Computer Mathematics

1 + 1 = 1

An Introduction to Boolean Algebra and Switching Circuits

ALSO BY WILLIAM R. PARKS

Beginning Algebra

Sets and Numbers

Algebraic Expressions

Linear Equations and Graphs

Properties of the Real Number System

Computer Mathematics

Sets and Flowcharts

Computer Number Bases

Introduction to Logic

1 + 1 = 1 An Introduction to Boolean Algebra and Switching Circuits

Statistics

Introduction to Statistics

Introduction to Gambling Theory

Computer Mathematics

1 + 1 = 1

An Introduction to Boolean Algebra and Switching Circuits

William R. Parks, B.S., M.S., Ed.M.

Hershey Books

www.wrparks.com

Cover

The four basic operations of arithmetic are displayed on the cover: Integer arithmetic and Boolean logic operations are performed in the arithmetic logic unit (ALU) of a computer.

Boolean algebra is named after George Boole (1815-1864), a mathematician and philosopher, who wrote the book, *The Laws of Thought,* which is the basis for Boolean algebra and logic used to design modern day computers.

George Boole (1815-1864)

www.wrparks.com

Hershey, PA 17033

ISBN: 978-0-88493-009-9

Library of Congress Control Number: 2014934440

INTRODUCTION

Boolean algebra is named after George Boole (1815-1864), a mathematician and philosopher who wrote, *The Laws of Thought*, which is the basis for Boolean algebra and logic used to design modern day computers.

In elementary algebra, 1 + 1 equals 2. In Boolean algebra, 1 + 1 equals 1.

Elementary algebra includes the study of variables and basic operations of arithmetic such as addition and multiplication. The variables can stand for a variety numbers such as the integers 1, 2, 3 ... or decimals numbers such as .5, 1.5, 2.7, etc. The basic rules of arithmetic apply in elementary algebra.

However, when we study Boolean algebra and switching circuits the values of variables are limited to two values denoted by the binary digits "1" and "0."

"1" can be represented in the physical world by a closed switch which will turn on a light and "0" can be represented by an open switch which will turn off a light.

The smallest unit of information in a digital computer is the binary digit which has only two possible states: 0 or 1. These two bits are the smallest units of information in a computer which can be stored or transmitted as high or low voltages representing 1 or 0.

Boolean algebra is directly related to the study of propositional logic where truth values are studied. The symbol "1" in Boolean algebra is considered "true" in propositional logic and the meaning of "0" is "false" in logic. For this reason, truth tables used in logic can be used to find the value of Boolean expressions.

Not all rules of elementary algebra work in Boolean algebra. Boolean algebra is not isomorphic (similar) to elementary algebra. However, Boolean algebra is isomorphic to logic.

In Boolean algebra and switching circuits there are three main operations that operate on binary variables that stand for either 1 or 0. These three operations are also found in the study of propositional logic and operate on logic variables that stand for either true or false.

In Boolean algebra the plus sign "+" means "OR" (called "disjunction" in logic) and the times sign ". " means "AND" (and called "conjunction" in logic). These two binary operations operate on two variables. However, the prime mark or tilde "~" means "NOT" (negation in logic) and it operates on only one variable at a time.

Conjunction in logic is denoted by an up arrow "∧" and as stated above stands for "AND." Disjunction in logic is denoted by a down arrow "∨" and stands for "OR." Negation, a unary operation, is often denoted by "~" (tilde symbol) and it stands for "NOT" in logic. In some books a bar is placed over a letter variable instead of using the tilde symbol.

Logic symbols will be used mainly in Volume 3, *Introduction to Logic,* which is part of this *Computer Mathematics Series*. Logic symbols are directly related to corresponding symbols in Boolean algebra.

Boolean algebra and logic have different symbols to represent the same concepts. Some books on Boolean algebra use corresponding logic symbols. This might confuse beginning students.

In this "Computer Mathematics Series" new information is presented in short sections with immediate testing. This form of personalized instruction is often used in on-line Internet based courses for distance learning. A small amount of information is presented in each section before advancing to the next section. Exercises are listed after several sections followed by an answer key.

INTRODUCTION TO BOOLEAN ALGEBRA AND SWITCHING CIRCUITS

1. In this study we will be concerned with switching circuits. A switch is either a mechanical or electrical device which can be either open or closed. When the switch is open, no electric current may pass through. When the switch is closed a current may pass through.

ILLUSTRATIONS:

open switch

closed switch

\- - - - - - - - - - - - -

2. Variables are used to represent switches because a switch's condition varies. It can be in one of two states, "0" corresponding to open and "1" corresponding to closed.

EXAMPLES:

(a) —— x —— (Switch x may be open or closed.)

If the above switch is open, then x = 0.

If the above switch is closed, then x = 1.

(b) —— y —— (Switch y may be either 0 or 1.)

"1" represents in the physical world a closed switch which can turn on a light and "0" represents an open switch which can turn off a light. In a computer they can be stored or transmitted as high or low voltages representing 1 or 0.

1 and 0 are called <u>binary digits.</u> Larger decimal numbers are stored as a series of binary digits. This topic is covered in *Computer Number Bases* in the four volume computer math series.

- - - - - - - - - - - - - -

3. Two switches x, y may be connected together either in <u>series</u> or in <u>parallel</u>.

EXAMPLES:

(a) —— x —— y —— (x and y are two switches in series)

(b)

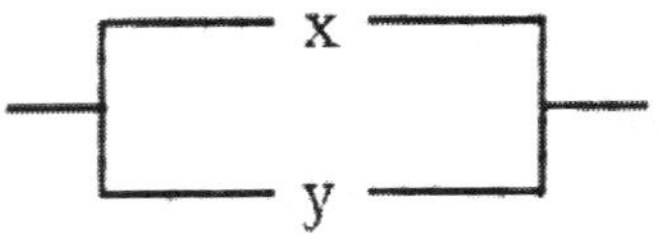

(x and y are two switches in parallel.)

4. There are four possible combinations of 0 and 1 for two switches x, y that can be configured together in series or in parallel as shown in the following table.

x	y
0 (open)	0 (open)
0 (open)	1 (closed)
1 (closed)	0 (open)
1 (closed)	1 (closed)

5. EXERCISE 1

1. A switch can be open or closed.

(a) True (b) False

2. Which of the following stand for switching states in Boolean algebra:

(a) 2 and 4 (b) 1 and 0 (c) 5 and 15

3. Variables (i.e. letter symbols) are used to represent switches because a switch can be in one of two states.

(a) True (b) False

4. How many different possible combinations are there for two switches x and y.

(a) One (b) Two (c) Three (d) Four

5. Two switches x and y may be connected:
(a) Only in series (b) Only in parallel (c) In series or in parallel

6. ANSWERS FOR EXERCISE 1

1. (a) True

2. (b) 1 and 0

3. (a) True

4. (d) Four: (0,0), (0,1), (1,0), (1,1)

5. (c) In series or in parallel

7. We denote two switches connected in parallel by the plus sign +.

EXAMPLE:

The expression "x + y" is represented by:

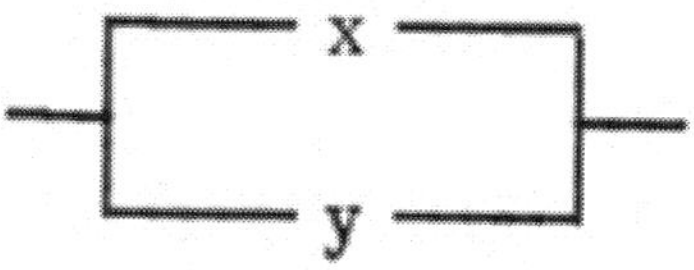

– – – – – – – – – – – – –

8. The below table representing x + y indicates when a <u>parallel circuit</u> is either open or closed. Such a display is called a "truth table." It shows a logic circuit output in the third column responding to various inputs in the first two columns. In logic 0 stands "false" and 1 stands for "true."

There are four possible combinations of 0 and 1 for two switches x and y. In this table we list the four possible combinations in the first two columns in the same order as presented in Section 4.

x	y	x + y	
0	0	0	(open)
0	1	1	(closed)
1	0	1	(closed)
1	1	1	(closed)

In the first horizontal line, switch x is zero. It is open. Switch y is also zero and it is open. Therefore, x + y must also be open and equal to zero.

An electrical current cannot pass through a parallel circuit x + y if both switches are open. However, in the next three rows, the current will flow if only one or both switches are 1 or closed.

Each row of this truth table can be represneted by the following Boolean equations:

$$0 + 0 = 0; \quad 0 + 1 = 1; \quad 1 + 0 = 1 \text{ and } \quad 1 + 1 = 1$$

RULE: The parallel circuit x + y in a truth table "equals" one, if at least one of the two switches "equals" one.

We can represent any line of a truth table by a circuit diagram. The below circuit diagram and the accompanying equation represents the first line in the table. The equation, 0 + 0 = 0, means "an open switch in parallel with an open switch is open." No current may pass through.

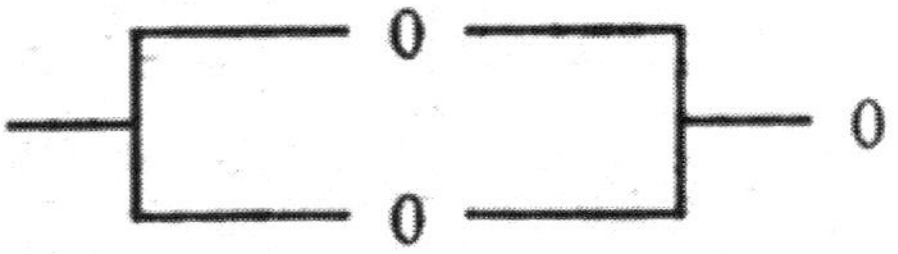

$$0 + 0 = 0$$

The below circuit diagram and the accompanying equation represents the second line in the table. The equation, 0 + 1 = 1, means “an open switch in parallel with a closed switch is closed.” A current may pass through such a circuit.

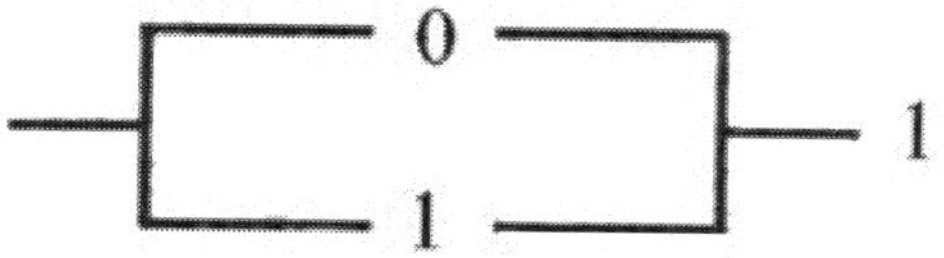

$0 + 1 = 1$

---

9. We denote two switches connected in series by a times sign “ • ” .

EXAMPLE:

The expression “x · y” represents:

The times sign means that the two switches x and y are connected in series. For electrical circuits this means that an electrical current is provided with only one path.

10. The following table for x • y indicates under which circumstances a <u>series circuit</u> is either open or closed.

x	y	$x \cdot y$	
0	0	0	(open)
0	1	0	(open)
1	0	0	(open)
1	1	1	(closed)

The first two columns in this table are the same as in the previous table in Section 8. However, in the third column we see a different set of binary digits. The only time a one or closed circuit condition occurs is when switches x and y are both one.

Rule: The series circuit x • y "equals" one only if both x and y are both one.

—— 1 —— 1 —— 1 $1 \cdot 1 = 1$

This circuit and the above equation represents the last line in the above series circuit table. The equation $1 \cdot 1 = 1$ means "a closed switch in series with a closed switch is closed. This circuit condition will allow current to flow and turn on a light bulb. Let's we assume with have a complete circuit with electricity flowing through the wire, the light bulb will be on.

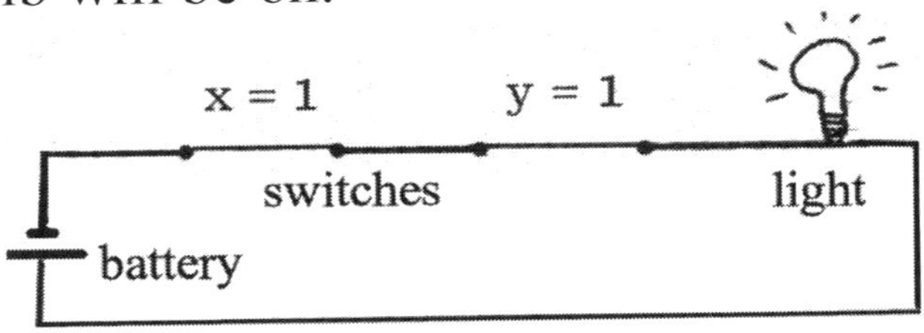

11. EXERCISE 2

1. We denote two switches connected in parallel by a:
(a) + sign (b) - sign (c) x sign (d) None of these

2. A parallel circuit x + y always equals one.

(a) True (b) False

3. A series circuit equals one when all switches equal one.

(a) True (b) False

4. If switch x is on and switch y is off in a parallel circuit then x + y = 1.

(a) True (b) False

12. ANSWERS FOR EXERCISE 2

1. (a) + sign

2. (b) False

3. (c) True

4. (a) True

13. Two switches in a circuit which are always in opposite states are called <u>complementary</u> switches. When the first switch is closed, then the second switch will be open.

In the below table switch x-prime represents the opposite of x. The prime mark is used to designate a switch in opposite state. The tilde mark "~" that is often used in logic can be interpreted to mean the negation of x. In this sense it is a <u>unary</u> operation operating on one switch. In previous examples we learned that the plus and times symbols are <u>binary</u> operations operating on two switches.

EXAMPLE:

x	x'
1	0
0	1

In the first line of the above table, x is 1 and x-prime is 0. Switch x is closed and switch x-prime its complement is open. In the next line switch x is 0 (or open) and switch x-prime (its complement) is 1 (or closed).

Let's examine the following two Boolean equations. The first equation is read, "x plus x-prime equals 1." This means that an electrical current will pass through a parallel circuit consisting of two complementary switches.

$$x + x' = 1$$
$$x \cdot x' = 0$$

The second equation, "x times x-prime equals 0." This means that an electrical circuit can not pass through a series circuit consisting of two complementary switches.

----

14. Series and parallel circuits can be combined to form more complex circuits.

EXAMPLES:

(a) Series-parallel circuit:

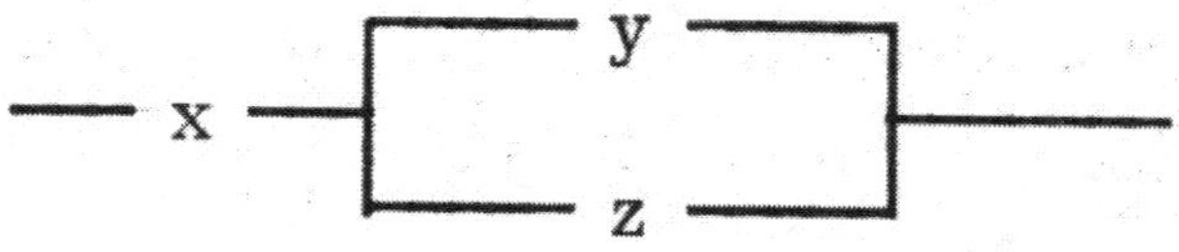

In the above circuit.switches y and z are in parallel and can be represented by $(y + z)$. This parallel circuit is in series with switch x.

The entire circuit is represented by the expression:

$$x \cdot (y + z)$$

(b) The circuit $x \cdot y + z$ is represented by a different configuration as follows:

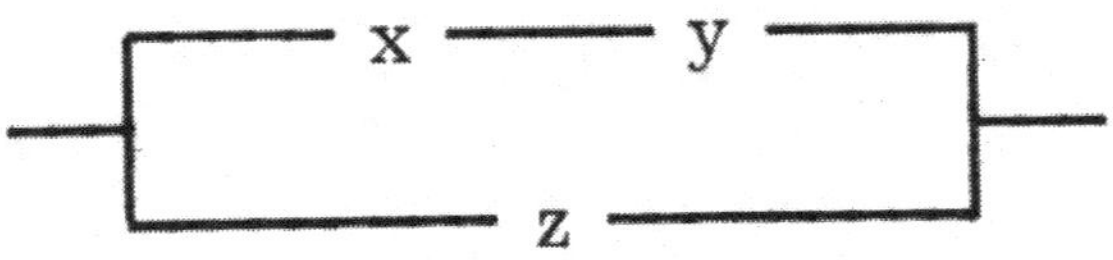

We interpret the expression x · y + z by deciding that x times y has precedence over y + z.

RULE: When more than one binary operator is used in a Boolean expression to represent a circuit, "times" is evaluated first and "plus" next. In logic, the unary operator NOT is evaluated first, then the binary operator AND which is the same as "times" in Boolean algebra. Finally OR is evaluated which is the same as a plus in Boolean algebra..

\- - - - - - - - - - - - -

15. The next table shows eight possible combinations when three switches are configured in the following series-parallel circuit:

$$x \cdot (y + z)$$

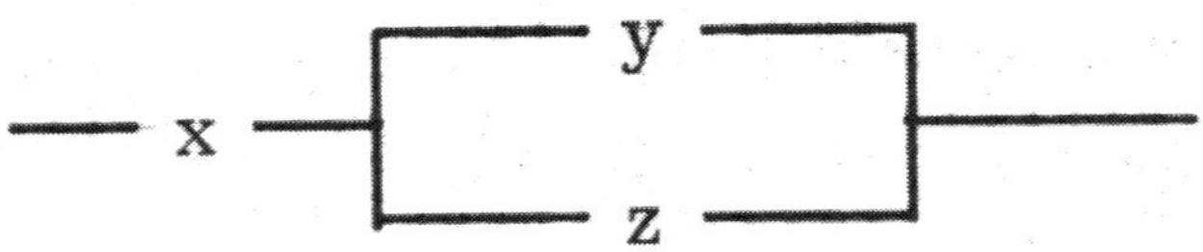

x	y	z	y + z	x · (y + z)
0	0	0	0	0
0	0	1	1	0
0	1	0	1	0
0	1	1	1	0
1	0	0	0	0
1	0	1	1	1
1	1	0	1	1
1	1	1	1	1

Here the corresponding set of values for y + z (a parallel circuit) are shown in the fourth column. The final results in the fifth column are calculated using the values in the first column in series with the values of the fourth column.

This table indicates clearly the exact order of "on" and "off" settings that will permit a current to flow through this three switch circuit. The combinations in the last three rows permit a current to pass through this particular configuration of switches. x has to be 1 in each case and at least one of either y or z switches must be 1.

In the above table, we placed zeros in the first row. This is arbitrary. In future tables the digits will be shown in reverse order. Three ones will be placed in the first row.

— — — — — — — — — — — —

16. EXERCISES 3

1. Complementary switches in a circuit are always in opposite states.

(a) True (b) False

2. Series and parallel circuits may not be combined.

(a) True (b) False

3. How many combinations of 0 and 1 are possible with three switches?

(a) 4 (b) 6 (c) 8 (d) 10

4. In the circuit $x \cdot (y + z)$, if $x = 1$, $y = 0$ and $z = 1$, the circuit will be

(a) "0" and will not pass current (b) "1" and will pass current

5. In the same circuit $x \cdot (y + z)$, if $x = 0$, $y = 1$ and $z = 1$, the circuit will be:

(a) "0" and will not pass current (b) "1" and will pass current

6. In the circuit $x \cdot y + z$, which binary operator has precedence and is evaluated first:

(a) $x \cdot y$ is evaluated first (b) $y + z$ is evaluated first

7. The circuit $x \cdot y + z$ is represented by the following diagram

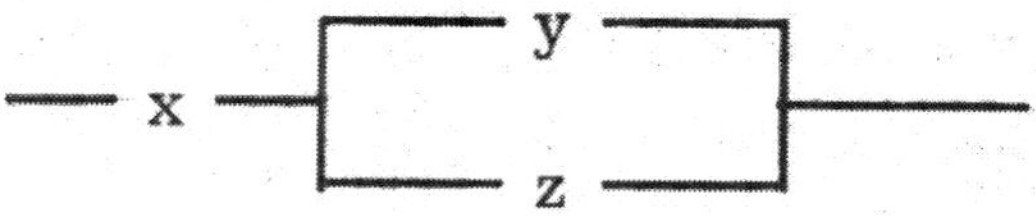

(a) True (b) False

17. ANSWERS FOR EXERCISE 3

1. (a) True

2. (b) False

3. (c) 8

4. (b) "1" and will pass current

5. (a) "0" and will not pass current

6. (a) $x \cdot y$ is evaluated first

7. (b) False. The circuit shown is: $x \cdot (y + z)$.

18. One purpose of Boolean algebra tables is to show that circuits are <u>equivalent</u>. If two columns representing different circuits in a given table are identical, then we can say that their circuits are equivalent.

EXAMPLES:

(a) The following table shows $x \cdot y + x$ is equivalent to x because the 0 and 1 entries are identical for each column. Electrical current will flow in both circuits as indicated in the first two rows. Note that we inverted the sets of ones and zeros from the way they appeared in previous tables. The order is arbitrary and will appear both ways in different textbooks. All possible combinations are still being displayed in the table.

x	y	$x \cdot y$	$x \cdot y + x$	x
1	1	1	1	1
1	0	0	1	1
0	1	0	0	0
0	0	0	0	0

(b) The next table shows $x \cdot (y + z) = x \cdot y + x \cdot z$
The equivalence of these two circuits is called the <u>first distributive property</u>. This property is also used in elementary algebra (x is distributed over y and z).

x	y	z	y + z	x · (y + z)	x · y	x · z	x · y + x · z
1	1	1	1	1	1	1	1
1	1	0	1	1	1	0	1
1	0	1	1	1	0	1	1
1	0	0	0	0	0	0	0
0	1	1	1	0	0	0	0
0	1	0	1	0	0	0	0
0	0	1	1	0	0	0	0
0	0	0	0	0	0	0	0

The fifth column of the binary digits ones and zeros (“on” and “off”) is identical in sequence to the binary digits in the last column. Therefore, the two circuits are equivalent.

The two distributive property expressions for these two circuits are shown in the following circuit diagrams. The two circuit diagrams appear to be different, however they are equivalent as shown in the previous table.

$$x \cdot (y + z) \quad = \quad x \cdot y + x \cdot z$$

— — — — — — — — — — — — —

19. The two sides of the below circuit equation appear to be different, however they are equivalent because each side produces the same table values. The equivalence of these two circuits is called the <u>second distributive property.</u>
The second distributive property of Boolean algebra is not allowed in elementary algebra. "x +" is distributed over y and z in Boolean algebra.

$$x + (y \cdot z) = (x + y) \cdot (x + z)$$

x	y	z	y · z	x + (y · z)	x + y	x + z	(x + y) · (x + z)
1	1	1	1	1	1	1	1
1	1	0	0	1	1	1	1
1	0	1	0	1	1	1	1
1	0	0	0	1	1	1	1
0	1	1	1	1	1	1	1
0	1	0	0	0	1	0	0
0	0	1	0	0	0	1	0
0	0	0	0	0	0	0	0

The fifth column of binary digits of ones and zeros ("on" and "offs") is identical to the list of binary digits in the last column. Therefore, the two circuits are equivalent. It's okay to distribute the "x+" in Boolean algebra..

\- - - - - - - - - - - - -

20. The theory of switching circuits has many applications. Electronic calculators, computers, and telephone systems are designed using the basic concepts of Boolean algebra and switching circuits.

EXAMPLE PROBLEM:
Design a circuit that will turn on a signal light when a majority of a committee of three persons vote yes. Each member of the committee will push a button to vote yes.

SOLUTION:
To record a secret majority vote, the following switching circuit can be used.

Let x, y, z represent the switches or "voting buttons."

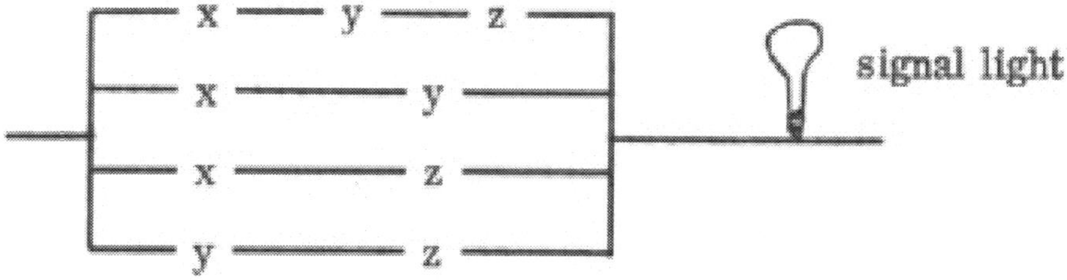

The above circuit will work to indicate a majority vote by turning on the light. However, the above circuit can be simplified. We can design a circuit that employs fewer switches but will be equivalent to the above larger circuit. Simplifying circuits can save money utilizing fewer switches on a circuit board that is part of an electronic device.

One method to simplify circuits in Boolean algebra is to find a simpler equivalent Boolean expression. This can be accomplished by using properties of Boolean algebra.

For example, we know that the following Boolean expression **x + x'** will always produce the same digital output, namely 1. Therefore we can substitute a single wire with no switching required in place of two switches in series that are complementary switches.

$$x + x' = 1$$

In fact, one of Boolean algebra's most important uses is in simplyfying digital circuits required for operating digital devices.

21. Let's simplify the circuit we used to record a simple majority vote of three persons. We construct tables for two equivalent Boolean expressions. The second Boolean expression in the second table is simpler and contains 3 less switches. The second circuit with 3 less switches produces the same output as the first circuit table but with less expense spent unecessary additional switches

I. (x · y) + (x · z) + (y · z) + (x · y · z)

x	y	z	(x · y) +	(x · z) +	(y · z) +	(x · y · z)	value:
1	1	1	1	1	1	1	1
1	1	0	1	0	0	0	1
1	0	1	0	1	0	0	1
1	0	0	0	0	0	0	0
0	1	1	0	0	1	0	1
0	1	0	0	0	0	0	0
0	0	1	0	0	0	0	0
0	0	0	0	0	0	0	0

II. This circuit has 3 less switches: (x · y) + (x · z) + (y · z)

x	y	z	(x · y) +	(x · z) +	(y · z)	values are the same:
1	1	1	1	1	1	1
1	1	0	1	0	0	1
1	0	1	0	1	0	1
1	0	0	0	0	0	0
0	1	1	0	0	1	1
0	1	0	0	0	0	0
0	0	1	0	0	0	0
0	0	0	0	0	0	0

Here is a simpler circuit diagram with 3 less switches that lights the signal light when a majority vote is cast.

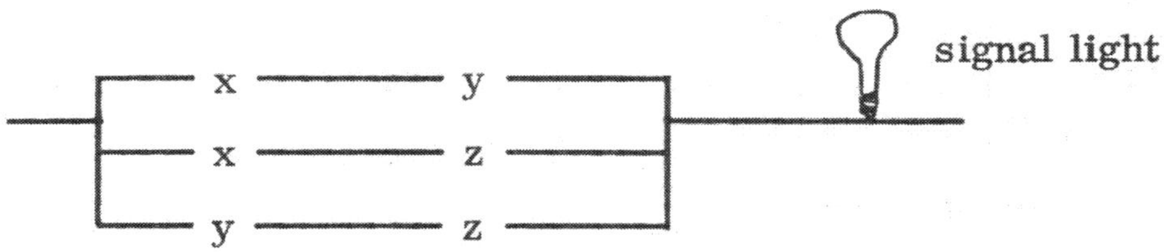

22. EXERCISE 4

1. If two columns representing circuits in a given table are identical, then we can say that their circuits are equivalent.

(a) True (b) False

2. If $x = 1$, $y = 0$ and $z = 0$, then what is the value of the circuit $x \cdot (y + z)$.

(a) 0 (b) 1 (c) none of these

3. The equivalence of the two circuits in the equation $x \cdot (y + z) = x \cdot y + x \cdot z$ is called the first distributive property.

(a) True (b) False

4. The first distributive property is also used in elementary algebra.

(a) True (b) False

5. The equivalence of the two circuits in the equation $x + (y \cdot z) = (x + y) \cdot (y + z)$ is called the second distributive property.

(a) True (b) False

6. The second distributive property is also used in elementary algebra.

(a) True (b) False

7. If $x = 1$, $y = 0$ and $z = 1$ then what is the value of $(x \cdot y) + (x \cdot z) + (y \cdot z) + (x \cdot y \cdot z)$

(a) 1 (b) 0

8. Construct a Boolean algebra table for the second distributive property: $x + (y \cdot z) = (x + y) \cdot (x + z)$.

23. ANSWERS TO EXERCISE 4

1. (a) True

2. (a) 0

3. (a) True

4. (a) True

5. (a) True

6. (b) False

7. (a) 1

8. The fifth column is the same as the last column, therefore, the second distributive property is true.

$$x + (y \cdot z) = (x + y) \cdot (x + z)$$

x	y	z	$y \cdot z$	$x + (y \cdot z)$	$x + y$	$x + z$	$(x + y) \cdot (x + z)$
1	1	1	1	1	1	1	1
1	1	0	0	1	1	1	1
1	0	1	0	1	1	1	1
1	0	0	0	1	1	1	1
0	1	1	1	1	1	1	1
0	1	0	0	0	1	0	0
0	0	1	0	0	0	1	0
0	0	0	0	0	0	0	0

24. Here is a summary of properties in Boolean algebra and switching circuits. Some but not all of these properties are found in elementary algebra.

I. Commutative properties of a sum and a product where changing the order of two switches does not change the results. The left side is equivalent to the right side of each of the following Boolean equations.

(a) Commutative property of a sum: $x + y = y + x$

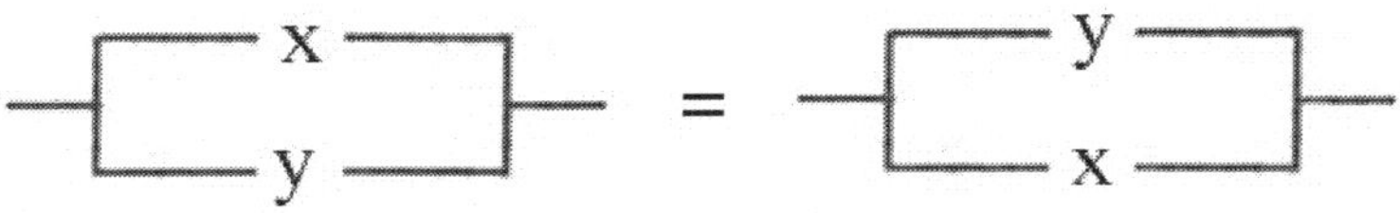

If $x = 1$ and $y = 0$, then $1 + 0 = 0 + 1$

The final result is $1 = 1$

(b) Commutative property of a product: $x \cdot y = y \cdot x$

—— x —— y —— = —— y —— x ——

If $x = 1$ and $y = 0$, then $1 \cdot 0 = 0 \cdot 1$
then the final result is $0 = 0$

II. Associative properties where changing the grouping does not change the results. The left side is equivalent to the right side of each of the following Boolean equations.

(a) Associative property of a sum:

$$(x + y) + z = x + (y + z)$$

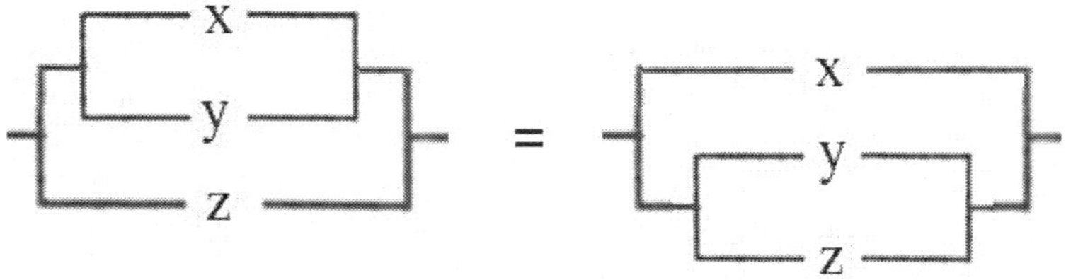

(b) Associative property of a product:

$$(x \cdot y) \cdot z = x \cdot (y \cdot z)$$

—— x —— y —— z —— = —— x —— y —— z ——

III. The identity element for + is 0 and the identity element for · is 1.

$$x + 0 = x$$
$$x \cdot 1 = x$$

IV. The first of the next two other Boolean identity properties is not valid in elementary algebra.

$$x + 1 = 1$$
$$x \cdot 0 = 0$$

V. Idempotent properties:

$$x + x = x \quad \text{(e.g. } 1 + 1 = 1 \text{ and } 0 + 0 = 0\text{)}$$
$$x \cdot x = x \quad \text{(e.g. } 1 \cdot 1 = 1 \text{ and } 0 \cdot 0 = 0\text{)}$$

VI. The following complementary properties were discussed in Section 13.

$$x + x' = 1 \quad \text{(e.g. } 1 + 0 = 1\text{)}$$
$$x \cdot x' = 0 \quad \text{(e.g. } 1 \cdot 0 = 0\text{)}$$

VII. The following distributive properties were discussed in Section 18.

$$x \cdot (y + z) = x \cdot y + x \cdot z$$

$$x + (y \cdot z) = (x + y) \cdot (y + z)$$

VIII. DeMorgan properties. In Exercise 5 you will be asked to verify DeMorgan properties by constructing a table of binary digits for each side of the following equations to verify that they produce or evaluate to an identical set of binary digits.

$$(x + y)' = x' \cdot y'$$

$$(x \cdot y)' = x' + y'$$

25. EXERCISE 5

1. Which property justifies the following Boolean equation?

$$(x + 1) + y = (1 + x) + y$$

(a) commutative property (b) associative property

2. Which property justifies the following Boolean equation.

$$(x + 1) + y = x + (1 + y)$$

(a) commutative property (b) associative property

3. Which property justifies the following Boolean equation?

$$(x \cdot 1) + y = (1 \cdot x) + y$$

(a) commutative property (b) associative property

4. Which properties justifies the following Boolean equation?

$$(x \cdot 1) \cdot y = x \cdot y$$

(a) commutative property (b) Identity property

5. Which property justifies the following Boolean equation.

$$(x + 0) + y = x + y$$

(a) Associative (b) Commutative (c) Identity

6.. Which property justifies the following Boolean equation.

$$(x \cdot 1) + y = x + y$$

(a) Associative (b) Commutative (c) Identity

7. Construct a table verifying: $(x \cdot 1) + y = (x + y)$

8. This time without giving you an answer in the below answer key, construct a table like we did in number 7 to verify that: $(x + 0) + y = x + y$. Each side of this equation on the table should have the same set or order of binary digits.

9. Construct a table for each of the following equations depicting DeMorgan's laws and verify that both sides of each equation are equivalent.

$$(x + y)' = x' \cdot y'$$

$$(x \cdot y)' = x' + y'$$

26. ANSWERS TO EXERCISE 5

1. (a) Commutative property of a sum.

2. (b) Associative property of a sum

3. (a) Commutative property of a product

4. (b) Identity property of a product.

5. (c) Identity property of a sum

6. (c) Identity property of a product

7. The fourth and fifth columns in the following table are the same therefore the two sides of the equation represent equivalent circuits and the following Boolean equation is true:

$$(x \cdot 1) + y = (x + y)$$

x	y	$(x \cdot 1)$	$(x \cdot 1) + y$	$(x + y)$
1	1	1	1	1
1	0	1	1	1
0	1	0	1	1
0	0	0	0	0

8. Each side of the equation in the table you constructed should have the same set and order of binary digits.

9. Verification of DeMorgan's first law which shows that the following Boolean equation is true.:

$$(x + y)' = x' \cdot y'$$

The fourth column (the left side) is equivalent to the last column (the right side). This verifies DeMorgan's second law.

x	y	x + y	(x + y)'	x'	y'	x' · y'
1	1	1	0	0	0	0
1	0	1	0	0	1	0
0	1	1	0	1	0	0
0	0	0	1	1	1	1

\- - - - - - - - - - - -

27. Using properties of Boolean algebra listed in this text, we can justify steps that are used to simplify digital circuits.

EXAMPLES:

(a) Simplify the expression xy + xy' Assume each term is a product like in elementary algebra where the times sign is missing. Name the property used in each step of the simplification process:

$xy + xy'$	$= x(y + y')$	First distributive property
	$= x \cdot 1$	Complimentary property
	$= x$	Identity property of a product

(b) Simplify the following expression $(x \cdot y) + (x \cdot y')$ similar to (a) above. In this case the times sign is shown explicitly.

$(x \cdot y) + (x \cdot y')$	$= x \cdot (y + y')$	First distributive property
	$= x \cdot 1$	Complimentary property
	$= x$	Identity property

(c) Prove that the following Boolean equation is true:

$$x \cdot y + x = x$$

In this proof, the below steps are taking place on the left side of the above equation and they are undergoing step-by-step change (justified by Boolean algebra properties) until the left side appears identical to the expression on the right side.

$x \cdot y + x \cdot 1$	Identity property
$x \cdot (y + 1)$	Distributive property
$x(1)$	Identity property
x	Identity property

28. EXERCISE 6

1. Name the property used in each step to prove the following equation's left side is equal to the right side: $(x + x \cdot y) = x$

(a) $x \cdot (1 + y)$

(b) $x \cdot 1$

(c) x

2. Prove that the following Boolean equation is true. Name each property that is used in your proof

$$x \cdot (x + y) = x$$

(3 – 9) Name the Boolean properties used to justify each of the following Boolean statements:

3. $x \cdot x = x$

4. $x + (y + z) = x + (z + y)$

5. $(x + 1) + y = x + (1 + y)$

6. $x = x \cdot 1$

7. $x' \cdot x = 0$

8. $x' + y' = (x \cdot y)'$

9. $1 + x = 1$

10. $x + (y \cdot z) = (x + y) \cdot (x + z)$

29. ANSWERS TO EXERCISE 6

1. (a) First distributive property

   (b) Identity property

   (c) Identity property

2. In this proof, the left side undergoes step-by-step change (justified by Boolean algebra properties) until it appears identical to the expression on the right side.

$$x \cdot (x + y) = x$$

$x \cdot x + x \cdot y$	Distributive property
$x + x \cdot y$	Idempotent property
$x \cdot (1 + y)$	Distributive property
$x \cdot 1$	Identity property
x	Identity property

3. Idempotent property

4. Commutative property

5. Associative property

6. Identity property

7. Complementary property

8. DeMorgan's property

9. Identity property

10. Second distributive property

This concludes *Introduction to Boolean Algebra and Switching Circuits.* Additional titles in the *Computer Mathematics Series* are: *Sets and Flowcharts, Computer Number Bases* and *Introduction to Logic.*

Made in the USA
Monee, IL
04 November 2026

67377174R00026